Tuart
Dwellers

Jan Ramage

Illustrations by Ellen Hickman

Department of
Environment and Conservation

Our environment, our future

At dawn, light flickers, as shadows play on the forest floor.

2

The mighty tuart towers above the woodland, where rough tracks part the undergrowth. Water-stains streak the trunk, and strips of bark hide little secrets.

A mudlark calls in a sharp, quick song, demanding to be heard. And a warbler flutes in the new day.

Out pops a hairy head. It's a mad hatterpillar, wearing a shiny black hat.

3

Little skink appears—too early—and slips back into safety, while dragon steps out for a snack.

The butcherbird calls his mate, and together they fill the stillness with their melody.

Babies trot, jumping, calling, squeaking, squawking—pleading to be fed.

Long-legged spiders tuck neatly away, while little hatterpillar crawls back out of sight.

The sun moves higher and the earth warms, breathing life into sleeping cicadas. They start their chorus, while parrots fill the sky with noisy chatter.

Light sparkles, leaves dance and seedpods pop.

Leaf litter crackles as tiny creatures move beneath.

Out slips a gecko. All is clear, and she stays to warm herself. Camouflaged against the bark, she almost disappears.

Hatterpillar peeps out, looking for a safe place to shed his old coat. And a bullseye borer buries deep into the wood.

Grass-skippers flit, while a hairymary trails
across the rough bark. Spitfires gather in huge
wriggling clumps near the trunk.

Little creatures busily prepare for the day—
building homes, hunting or hiding. A tiny crab
spider folds over a leaf and waits. Behind his twig,
mad hatterpillar once again peeps out.

Treecreepers grip onto the flaky trunk and peck.
Inspecting every crevice, their sharp eyes miss little.
They hop from branch to branch, darting this way
and that, hanging and probing for honey nectar.

Mad hatterpillar stays very still, and a jewel
beetle flicks her wings to escape.

Sunlight filters down, dappling patterns in the
undergrowth, where mantis stands on guard and
ready to strike.

Noisy black cockatoos arrive to announce the coming rain. They ribbon the sky with outstretched wings.

Hunting for grubs and insects, they rip off pieces of bark and break new shoots, before flocking away. Clouds gather. Insects buzz and ants scurry, hurrying to gather supplies before the wet. And a johnny hairylegs scuttles out of sight.

A loud crack fills the sky, and thunder rolls across the valley.

But tucked inside leaves or buried deep underground, other creatures continue to sleep.

Little hatterpillar is busy. Stretching and wriggling, he splits open his coat and slowly bulges out. Easing free, he slips off his old capsule and his skin falls away.

Lightning strikes—a blinding flash—a thundering crash! A limb falls.

Swollen clouds burst. And great sheets of rain drench the undergrowth. Water tracks down in little rivers, making puddles.

The roots of the tuart spread, stretching deep into the earth to drink. Branches are washed by the rain. The tree breathes; its girth expands.

And hatterpillar rests under a leaf.

As quickly as it came, the storm goes. Sunlight streams through the canopy. It dances on the forest floor. And the sky explodes with colour.

Branches hang in curtains of green. All is still.

16

Little hatterpillar has another cap to add to his hat. He twists and wriggles, gathering it up, and finally the task is done. His pointed hat stands tall, with four shining beads stacked one on top of the other.

A triller appears. Spotting the shiny hat through the branches, she dives down to snatch it—and as quick as a flick, it's gone.

A rainbow bee-eater collects his supper before dusk. Darting in and out, he disappears into his secret burrow.

A jezebel alights. The woodland sparkles. And little frog sits alone.

17

Shadows grow and spread into night.
The bush becomes dark.

The lonely call of the boobook breaks
the stillness and echoes through the forest.
He waits and watches. Other creatures
emerge to begin their nightly feast.

Hatterpillar wriggles and munches. His
new skin stretches and grows as he
prepares for his transformation.

Treetops whisper as the night watch awakes.

Hatterpillar hides under scraggy twigs and leaves. He weaves his cocoon from silken thread, adding small pieces of bark, twigs and frass. In time, he will emerge as a delicate moth.

Below, among the leaf litter, a tiny tuart tree begins to grow. Soon new leaves will unfurl as fine roots reach into the damp earth.

And the cycle of life continues in the tuart forest.

Fascinating Facts about the Habits and Environment of the Tuart Dwellers

Tuart longhorn beetles lay their eggs on bark in the rough areas of small branches.

Tuart longhorn grubs (larvae) soon develop and bore into the tree. They can do a lot of damage to tuarts, especially those stressed by disease or lack of water.

Kino (red sap) on the bark is a sign that the tuart longhorn grub or another wood-boring grub has made its home in the tree.

The **parasitic wasp** helps the tuart tree by hunting the larvae of the tuart longhorn. Using its long ovipositor (egg-laying tube), it lays its eggs into the body of the longhorn grub, eventually killing it. The egg develops into a young wasp grub (larva), which hatches from the body of the longhorn and goes on to develop into an adult wasp.

See pages 2–3

Bracket fungi grow like big cushions on trees such as the tuart, providing food for many types of insects.

Galls growing from the branches or on the leaves indicate that an insect has visited the tree. These swellings on a tuart leaf are from a male scale insect.

The **mudlark** is territorial and in nesting season will attack other birds, including the larger magpie and even its own reflection. It has the unique ability to fly almost vertically from a standing position on the ground.

During breeding season, **magpie** males warble their beautiful call all through the night.

The **mad hatterpillar** is an extremely hairy caterpillar. Many predators are deterred by its long, thick hair. It also has a clever "hat trick". When it grows and needs to shed its old skin, it attaches the old capsules onto the top of its head, stacking each new one on top of the others, like a tall hat. If a bird comes along, it is hoped it will spot the "hat" part sticking out and eat that instead of the caterpillar. The hatterpillar adds to its "hat" each time it needs a change of skin.

When being chased, the **skink** gives its tail a quick twist. This is a "one-off" trick, which breaks the backbone, allowing the tail end to drop off. Sometimes two tails will grow back to replace the lost one, but the skink can lose its tail only once in its lifetime.

To frighten away predators, the **western bearded dragon** flattens its body and stretches up to stand. It opens its mouth wide to show a bright yellow lining and puffs out the spines around the bottom jaw to look fierce.

Grey **butcherbirds** are remarkable songsters. Related to the magpie, they are also territorial. They often use a sharp stake to hook their meal before eating it, or store their prey in the fork of a tree for later eating. This puts them in the rare category of "tool-user" birds.

Magpie babies are protected, fed and cared for by the members of their group. Magpies are unusual as they live in groups of up to twenty-four. Each group has its permanent territory.

Huntsman spiders like to hide under bark. They jump away to escape predators.

See pages 4–5

Cicadas lay their eggs in cracks in tree trunks and branches. These noisy insects are often called "tick-tocks" or "sand grinders". Their shells are left behind when they emerge into beautiful winged adults.

Twenty-eight parrots have a call that sounds just like they are saying "twenty-eight". They nest in tree hollows and are a common sight in the tuart forest.

Rainbow lorikeets are an introduced species. Being more aggressive, lorikeets often win in a fight with twenty-eights over nesting territory.

Pink and grey galahs are fast fliers, reaching speeds of 50 kilometres per hour. They can be seen in flocks of many hundreds, although they were rare when European settlers arrived in Western Australia. They usually mate for life and can live to a great age.

Marbled geckos have very fine, pointed hairs on their feet, which help them cling to trees. They can't blink their large eyes, but instead use their tongues to wipe them clean. Geckos are the only "talking" reptile and make a loud "squerk" sound to frighten away predators.

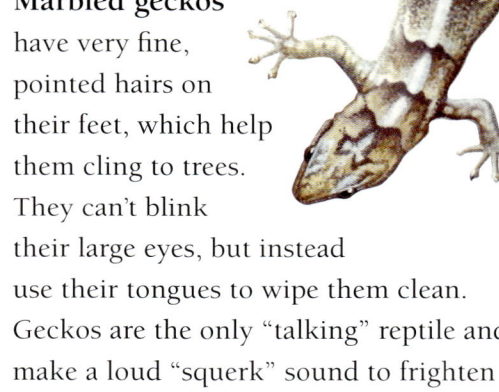

Bullseye borers eat into the wood of trees, just underneath the bark. If you notice sap or sawdust running down a branch or trunk, look closer and you might see a "bullseye" hole, which shows that a borer is living there.

See pages 6–7

Wedge grass-skippers flit from flower to flower. They lay their eggs on spear-grass, which is common under tuart. Adult butterflies emerge in autumn rather than spring.

Spitfires (sawfly larvae) spit a bitter liquid at predators and over themselves, which makes them taste disgusting. To confuse predators, they cluster in huge repulsive clumps.

Sawfly females do not mate with a male; they simply *clone* themselves. The female cuts into the leaf of a tuart using a saw-like device on her abdomen, and lays her eggs in the slit. She protects the eggs until they are ready to hatch; if disturbed, she buzzes her wings angrily and opens her jaws to scare away predators.

Hairymary caterpillars have very hairy bodies, which deter predators. The bright red stripe acts as a warning.

Twisted moths look just like a dry, twisted leaf as they lie well hidden on the ground, among dead leaves.

The **caterpillar** of the **twisted moth** looks like a new stem of a branch. It holds on tightly, using thin strands of silk, and feeds on tuart foliage.

Crab spiders build homes in folded leaves. They stick the edges together and hide inside. When an unsuspecting bug walks past, the spider pounces and then pulls it into its home to dine in safety.

Antlion nymphs build sand traps that look like moon craters. Lying at the base the nymph waits. When an ant comes along, the nymph excitedly tosses up more and more sand, making sure the ant loses its footing. It then grabs the ant and sinks into the sand to eat its prey.

Stick insects are the largest insects in the world—some as long as 20 centimetres. They look just like a twig on a tuart branch.

See pages 8–9

25

The **rufous treecreeper** was once common in the tuart forests. In what is often called the "funeral march", the bird starts at ground level on the tree trunk and marches up, looking for insects hidden in the bark.

Singing honeyeaters can be seen hanging in strange positions, probing blossoms for insects or nectar. They can also be spotted gripping tree trunks, inspecting the bark for a tasty snack of larvae or raiding smaller bird nests to feed on the eggs.

The **jewel beetle** will drop suddenly like a seedpod to hide in the leaf litter, or, with a quick click of its wings, will eject itself away from danger before flying off. Some species simply play dead. The larvae live in hidden spots of the tree trunk and bore into the wood and root systems.

The small, nomadic **mistletoe bird** eats sweet berries from the mistletoe plant, a parasitic bush that grows on trees such as the tuart. It makes a unique nest, using spider webs, cocoons and fluffy seeds firmly bound together into a smooth, felt-like material. It is also a good mimic and sings the songs of other birds.

The **praying mantis** rocks backwards and forwards like a leaf swaying in a breeze, or stands completely still like a twig. Its huge eyes help it to see prey from all angles, even from behind.

See pages 10–11

Johnny hairylegs centipedes live under bark, often near the base of a tree such as a tuart.

This big centipede has a poisonous bite. Its tail end looks more like the head, so anything attacking the tail may get a rude shock.

The **tuart leaf miner** lives between the two outer skins of the tuart leaf and eats the softer tissue in between. It cuts the outer skins of the leaf and sticks them together to make a sac. This sac falls to the ground (with the caterpillar inside), leaving a hole in the leaf.

Red-legged weevils lay their eggs under the ground, where the nymphs attach themselves to the root system of a tree. They use the same method of escape as the jewel beetle—dropping like a seedpod into the leaf litter—and also play dead.

Tuart bud-weevils can cause lots of damage to a tuart tree. They lay one egg on each flower bud. When the grub hatches, it chews out the inside of the bud down to the base, preventing it from flowering or producing seeds.

Black cockatoos move around in large flocks in autumn and winter. The first settlers knew of this migration and it quickly became a signal that rain was on its way. These endangered birds will nest in tree hollows in large tuart trees and also feed on the nuts, always using their left foot. They hunt for insects and grubs under the bark, using their powerful beaks.

Bull ants are fearless and will attack a fully grown man to defend their territory. They stab with their back sting after holding their prey firmly with their front "jaws".

27

See pages 12–13

Buds begin as very tiny swellings attached to a strap.

As the white stamens on the flower grow, they push off the **bud-caps**. Bud-caps can be found scattered beneath a tuart tree in autumn when it is in flower.

The **flower** is fertilised by many different kinds of insects. Once fertilised, it develops into a small and woody **fruit**.

When the fruit falls, the valves gradually open, allowing the seeds to fall out. Strong winds and storms can carry branches with the fruits and seeds, scattering them far away from the adult tree. Ants will collect seeds and take them away to bury underground.

Seeds begin to develop roots after the autumn and winter rains. New roots, called "feeder roots", are no thicker than a thread. The tiny hairs attached to a young root take up water, minerals and air from the ground. The roots suck up water and minerals from the soil, too, and pump these up through the whole tree to help it grow. Roots breathe air in the tiny spaces between particles of soil. Overwatering can flood these pockets of air, causing a tree to drown.

The tree breathes over its whole surface. It gives off oxygen, which most creatures need to survive. Without trees, life on Earth would not exist.

See pages 14–15

White-winged trillers are highly nomadic birds that have a beautifully long trilling call. Caterpillars make up 80 per cent of their diet.

Jezebel butterfly larvae look like bird droppings—to birds and other predators. Butterflies do not spin cocoons like moths. Instead, the pupa becomes a chrysalis. It hangs by a silken halter (like a belt) around its middle. When the butterfly finally emerges from the chrysalis, its body is moist and soft. It has to pump blood from its body into its wings, which can take up to an hour.

Motorbike frogs have a call that sounds exactly like a roaring motorbike changing gears. Numbers are declining due to loss of habitat, water salinity (caused by clearing too many trees) and a fungus disease called chytrid (kit-rid), which affects the frog's skin.

Rainbow bee-eaters burrow into the ground for about a metre to make their nest, and lay four to five eggs on the sandy floor. They skilfully fly straight into the burrow, much like a guided missile, to avoid leaving "footprints" or any scent that might attract a hungry monitor lizard.

See pages 16–17

The **white-striped freetail bat** is endangered due to habitat loss. It lives in tree hollows and eats flying insects, and can reach speeds up to 30 kilometres per hour. It uses a "chit" call when approaching an insect. The call quickens to a feeding "chitter". The sound echoes from this noise help the bat to identify the target and check the position of the prey before it has a chance to escape.

These **shark-fin** marks show that a leaf-eating beetle or a weevil has visited the branch. Serrated marks are left behind when a caterpillar has visited for a snack. Some caterpillars, such as the mad hatterpillar, skeletonise leaves by eating off the surface.

The **boobook owl**'s distinctive "boo-book" call can be heard at night, but the owl is not often seen. It flies soundlessly, due to its soft, rounded feathers, which grow all the way down to its talons. An owl's eyes are big, four times larger than a human eye. It also has excellent hearing. One ear is always slightly higher than the other, which gives an above and below position for sound. This helps the owl to pinpoint the exact spot before it swoops silently to make a kill.

Brush-tailed phascogales are treetop-dwelling acrobats commonly called "wambengers". These endangered marsupials live in tree hollows. Though small in size, they are ferocious and have been known to kill domestic chickens. The males die shortly after mating and so the females are left alone to raise the young.

Slaters, **click beetles** and **scarabs** often appear at night, when it is safer for them to feed.

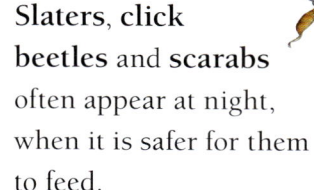

See pages 18–19

The common **brushtail possum** is omnivorous (eats plants and animals). Even if there are tree hollows for their homes, they will often choose to live in roof spaces and chimneys. Males use scent glands on their chests and also urine to mark their territory, which warns other possums away.

Western ringtail possums are herbivores. These nocturnal marsupials make a nest (called a drey) using twigs and, if possible, the soft leaves of the peppermint tree.

They sometimes use tree hollows in a tuart or the base of a balga grasstree to nest. Western ringtail possums are a threatened species, as land clearing has taken their natural habitats.

The **cocoon** of the mad hatterpillar is rarely seen, as these creatures are experts in camouflage. It is woven from fine silken thread, to which the hatterpillar attaches tiny twigs, shreds of bark, frass (droppings), some of its hair and its "hat" (capsule), if it is still wearing one. Cocoons may be hidden in leaf litter on the ground, or in small crevices in the bark of eucalypts such as the tuart, or among foliage attached to the fork of a twig. Inside the cocoon, the hatterpillar moults to become a pupa, and finally emerges as a moth.

Leaves of a **seedling** are broad to catch the sunlight, which helps it grow. In summer a seedling will tilt its leaf edges towards the sun's rays to avoid "sunburn".

Out of the hundreds of thousands, probably millions, of seeds produced by a tuart tree, only a few will become seedlings, and even fewer will grow into mature trees.

See pages 20–21

This book is dedicated to the tuart, a beautiful and generous species found in a small coastal strip in Western Australia between Lancelin and Busselton. The tuart is well known to people who live in and around Perth, where many families have, or once had, one growing in their gardens.

Unfortunately, tuart woodlands are dwindling, due to urban growth, repeated attacks by wood-boring beetles and the clearing of some of our forests through agriculture and mining. Along with the demise of these trees go the fascinating communities of insects, lizards, mammals and birds that are found in tuart woodlands. Awareness of the bushes, trees and shrubs that naturally complement the tuart—and replanting these species in and around our natural bushlands, as well as in our gardens—will help to replace these lost areas.

Special thanks to my father, Douglas Job—an environmentalist before his time—who planted a tiny tuart tree in our back garden, and to my mother, Lesley, for her inspiration and love. *J. R.*

To all those who encouraged, supported and inspired my artistic talent. *E. H.*

Acknowledgments

Thanks to Ralph and Phyllis Soderland, custodians of a heritage tuart tree in City Beach known as "The Happy Tree".

Many thanks to the Western Australian Government's Tuart Response Group for initiating this project; and to Alan Walker, Drew Haswell, Robert Powell and Ron Kawalilak, who have generously given their support and expertise in the development of *Tuart Dwellers*.

Thanks also to the following scientists/ technicians from the Western Australian Museum, the Department of Environment and Conservation, Murdoch University and The University of Western Australia: Tom Burbidge, Allan Wills (beetles and insects); Ron Johnstone (birds); Ric How, Brad Maryan (reptiles); Pauline Southgate (frogs); Jan Taylor (insects); Paul de Tores (possums); Tony Friend (brush-tailed phascogales); Mike Bamford (insects and birds); Norm McKenzie (bats); Peter Mawson (zoologist); Michael Hislop (botanist); Bernd Liebold (photography).

First published in 2008 by
Department of Environment and Conservation, Western Australia
Locked Bag 104,
Bentley Delivery Centre
Western Australia
www.dec.wa.gov.au
Reprinted in 2009

Text copyright © Department of Environment and Conservation and Jan Ramage 2008
Illustrations copyright © Ellen Hickman 2008

National Library of Australia Cataloguing-in-Publication entry:

Ramage, Jan, 1955– .

    Tuart dwellers.

    For children.

    ISBN 9781876615307 (hbk.).

    1. Forest insects—Western Australia—Juvenile literature. 2. Forest animals—Western Australia—Juvenile literature. 3. Eucalyptus—Ecology—Western Australia—Juvenile literature. 4. Forest ecology—Western Australia—Juvenile literature. I. Hickman, Ellen, 1968– . II. Western Australia. Department of Environment and Conservation. III. Title.

577.309941

Produced by Benchmark Publications Pty Ltd
Consultant editor Amanda Curtin
Designed by Sandra Nobes, Tou-Can Design, Melbourne
PrePress by Hell Colour, Melbourne
Printed by Everbest Printing Co., China

Jan Ramage is a teacher, writer and nature lover. She grew up with a tuart tree in her own back garden and has always been fascinated by the intelligence of our natural world. Her first book, *Eyes in the Night*, won a Crichton honour award in the CBC Awards for its illustrations by Laura Peterson, and was shortlisted for an environmental award by the Wilderness Society. *Tuart Dwellers* is Jan's second book.

Ellen Hickman lives in the coastal town of Albany, Western Australia, where she works as a botanist and artist with a love for the Australian bush. Her publications include the picture book (with Rina Foti) *Hooray for Chester!* Ellen's illustrations for *Tuart Dwellers* use watercolour, pencil and gouache, which reflect the subtle changes in light at play in the tuart forest.